page 1

page 20

page 2

page 4

page 6

page 3

page 21

page 25

page 9

page 27

page 8

page 7

page 28

More stickers at back of book!

Count Back

Count 10 to 1,
find something fun!

Connect the dots in order.
Start with **10.**

Color the picture.

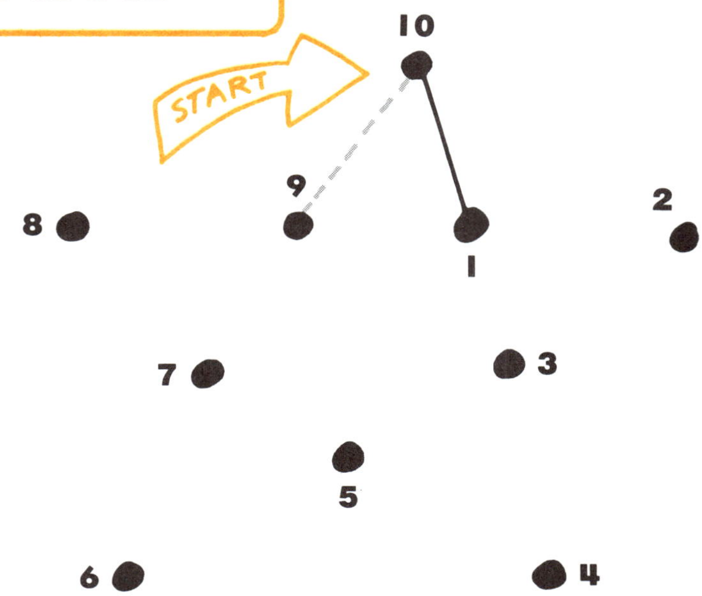

Find the stickers with the missing numbers.

Count 10 to 0,
be a hero!

Get Ready

Count how many are left.
Write the numbers.

How many are left? 2

How many are left? ______

How many are left? ______

Use a sticker to answer.

Look! Look! Here's the  .

Once 3 , now there's . . .

Skill: To build readiness for subtraction.

Sail Away

Write how many are left.

How many are left? $5 - 2 =$ 3

How many are left? $3 - 2 =$ ____

How many are left? $3 - 0 =$ ____

How many are left? $4 - 2 =$ ____

Use a sticker to answer.

How does the ocean say "Bye-Bye"?

Bye-Bye!

Write how many are left.

5 . 3 say "Bye-Bye." How many are left? 2

4 . 1 says "Bye-Bye." How many are left? ____

5 . 5 say "Bye-Bye." How many are left? ____

Use a sticker to answer.

What has 4 wheels and says "Oink"?

Skill: To subtract numbers 0–5 from numbers to 5.

Get Rolling

Subtract.

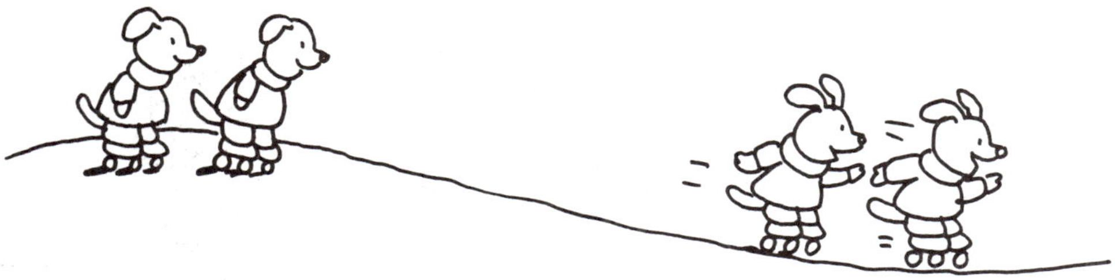

How many are left? $4 - 2 =$ 2

How many are left? $5 - 4 =$ ____

How many are left? $3 - 1 =$ ____

Use a sticker to answer.

What kind of dog is never cold?

Hats Off!

Subtract.

How many are left? 5 – 3 = 2

How many are left? 5 – 4 = ____

How many are left? 3 – 1 = ____

How many are left? 4 – 3 = ____

Use a sticker to answer.

What has a head and a foot, but no body?

Skill: To subtract numbers 0–5 from numbers to 5.

On Your Feet

Subtract.

X out 1.

How many are left?

4 − 1 = 3

X out 4.

How many are left?

5 − 4 = ____

X out 2.

How many are left?

2 − 2 = ____

X out 3.

How many are left?

4 − 3 = ____

Use a sticker to answer.

What has 2 hands but no feet?

Away They Go

Write how many are left.

How many are left? $7 - 3 =$ 4

$6 - 4 =$ ____

$8 - 5 =$ ____

$7 - 5 =$ ____

Use a sticker to answer.

What animal always goes to sleep in shoes?

Skill: To subtract numbers 0–8 from numbers to 8.

Take Off

Write how many are left.

How many are left? $8 - 1 = 7$

$8 - 2 = \_\_\_$

$7 - 2 = \_\_\_$

$8 - 7 = \_\_\_$

Use a sticker to answer.

6  **way up high.**

4  **land.**

How many in the sky?

In the Kitchen

Subtract.

How many are left? $8 - 4 =$ 4

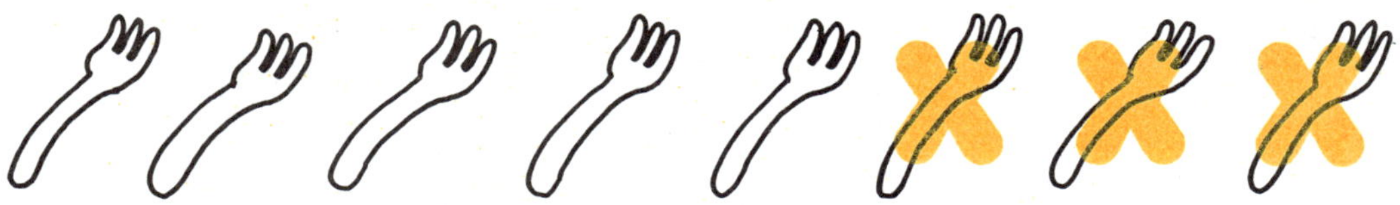

How many are left? $8 - 3 =$ ____

How many are left? $7 - 6 =$ ____

How many are left? $7 - 2 =$ ____

Use a sticker to answer.

7 of .

Oops! 5 spill.

How many are left?

Skill: To subtract numbers 0–8 from numbers to 8.

Keys, Please

Cross out and subtract.

X out 6.

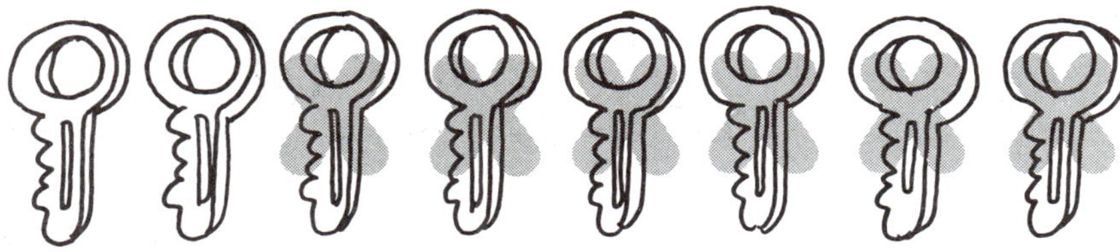

How many are left? $8 - 6 = 2$

X out 5.

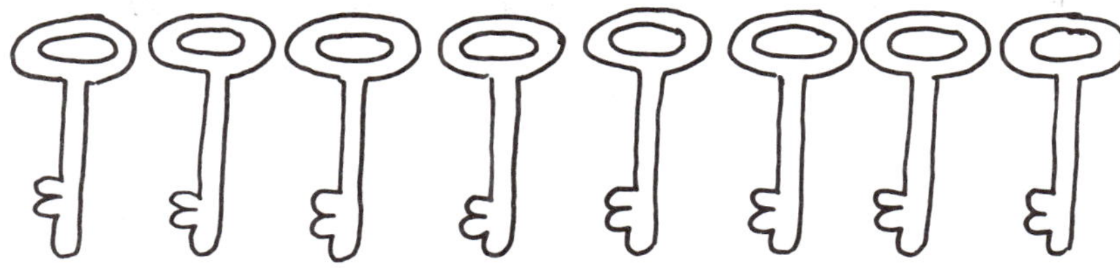

How many are left? $8 - 5 =$ ____

X out 3.

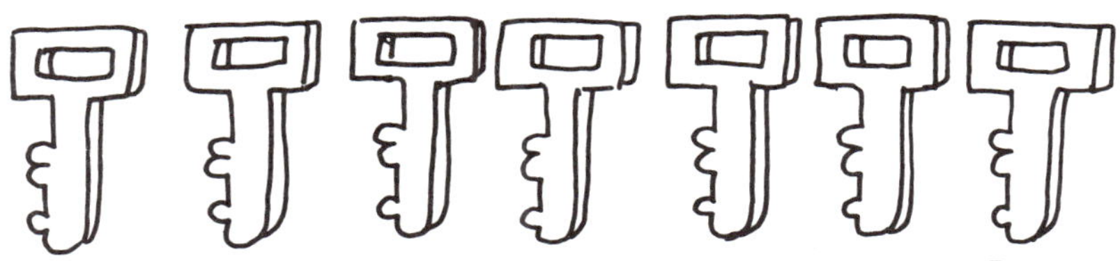

How many are left? $7 - 3 =$ ____

X out 6.

How many are left? $7 - 6 =$ ____

Use a sticker to answer.

What has many keys but no locks?

Top to Bottom

Subtract.

$$\begin{array}{r} 8 \\ -\ 5 \\ \hline 3 \end{array}$$

How many are left? 3

$$\begin{array}{r} 8 \\ -\ 7 \\ \hline \end{array}$$

How many are left? ____

$$\begin{array}{r} 8 \\ -\ 6 \\ \hline \end{array}$$

How many are left? ____

$$\begin{array}{r} 7 \\ -\ 5 \\ \hline \end{array}$$

How many are left? ____

$$\begin{array}{r} 7 \\ -\ 4 \\ \hline \end{array}$$

How many are left? ____

Use a sticker to answer.

What kind of key cannot unlock a door?

Skill: To subtract numbers through 8 in vertical form.

Wind It Up

Cross out and subtract.

$$\begin{array}{r} 6 \\ -\ 5 \\ \hline \end{array}$$

How many are left? ____

$$\begin{array}{r} 8 \\ -\ 3 \\ \hline \end{array}$$

How many are left? ____

$$\begin{array}{r} 7 \\ -\ 6 \\ \hline \end{array}$$

How many are left? ____

$$\begin{array}{r} 6 \\ -\ 6 \\ \hline \end{array}$$

How many are left? ____

$$\begin{array}{r} 8 \\ -\ 4 \\ \hline \end{array}$$

How many are left? ____

Use a sticker to answer.

$$\begin{array}{r} 8 \\ -\ 5 \\ \hline \end{array}$$

How many are left? ____

Toy Stories

Read the story.

Write the answer.

4  on the 

3  fall off. How many  are left? ____

---

8  on the 

5  fall off. How many  are left? ____

---

8  on the 

7  roll off. How many  are left? ____

---

6  on the 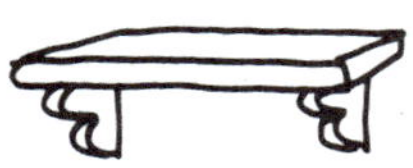

3  roll off. How many  are left? ____

Use a sticker to answer.

7 toy cars all in a row.
6 can go fast.
How many go slow?

Skill: To solve subtraction story problems.

When the farmer got to the apples his wife cried, "Why, this morning I could not spare one apple for our neighbor. Now I have a whole sack! How wonderful!"

So the two rich men had to pay $100 to the farmer for the sack of rotten apples.

And that, of course, was the very best trade of all!

What Can You Get for a Horse?

A farmer and his wife lived on a little farm with their horse.

One day the farmer said, "Let's sell the horse for something better."

"Fine with me," the wife said.

The farmer took the horse to the fair. On the way he met a man with a cow.

"Would you like to trade?" asked the man.

"Well," said the farmer, "my horse is worth more money. But a cow is better. It will give us milk."

And so they traded.

The men laughed. "We'll bet $100 to your sack of apples that your wife will be angry."

So off they went, and the farmer told the story all over again. His wife clapped at every trade!

Soon the farmer met two rich men. He told them about his trades.

"Why, you fool!" said the men. "Your wife will be very angry."

"I don't think so," said the farmer.

Next the farmer met a man with a sheep.

"Would you like to trade?" asked the man.

"Well," said the farmer, "my cow is worth more money. But a sheep is better. It will give us wool to keep us warm."

And so they traded.

Then a woman with a goose came along.

"Would you like to trade?" asked the woman.

"Well," said the farmer, "my sheep is worth more money. But a goose is better. It will give us beautiful feathers."

And so they traded.

$9 – $7 = $2 lost

Next the farmer met a man with a sack of rotten apples.

"Would you like to trade?" asked the man.

"Well," said the farmer, "my goose is worth more money. But my wife would like those better. Last year we had only one rotten apple on our tree."

And so they traded.

$7 – $2 = $5 lost

What Is It?

Subtract.

Color the spaces to match the answers.

2= red 4= green

3= blue 5= yellow

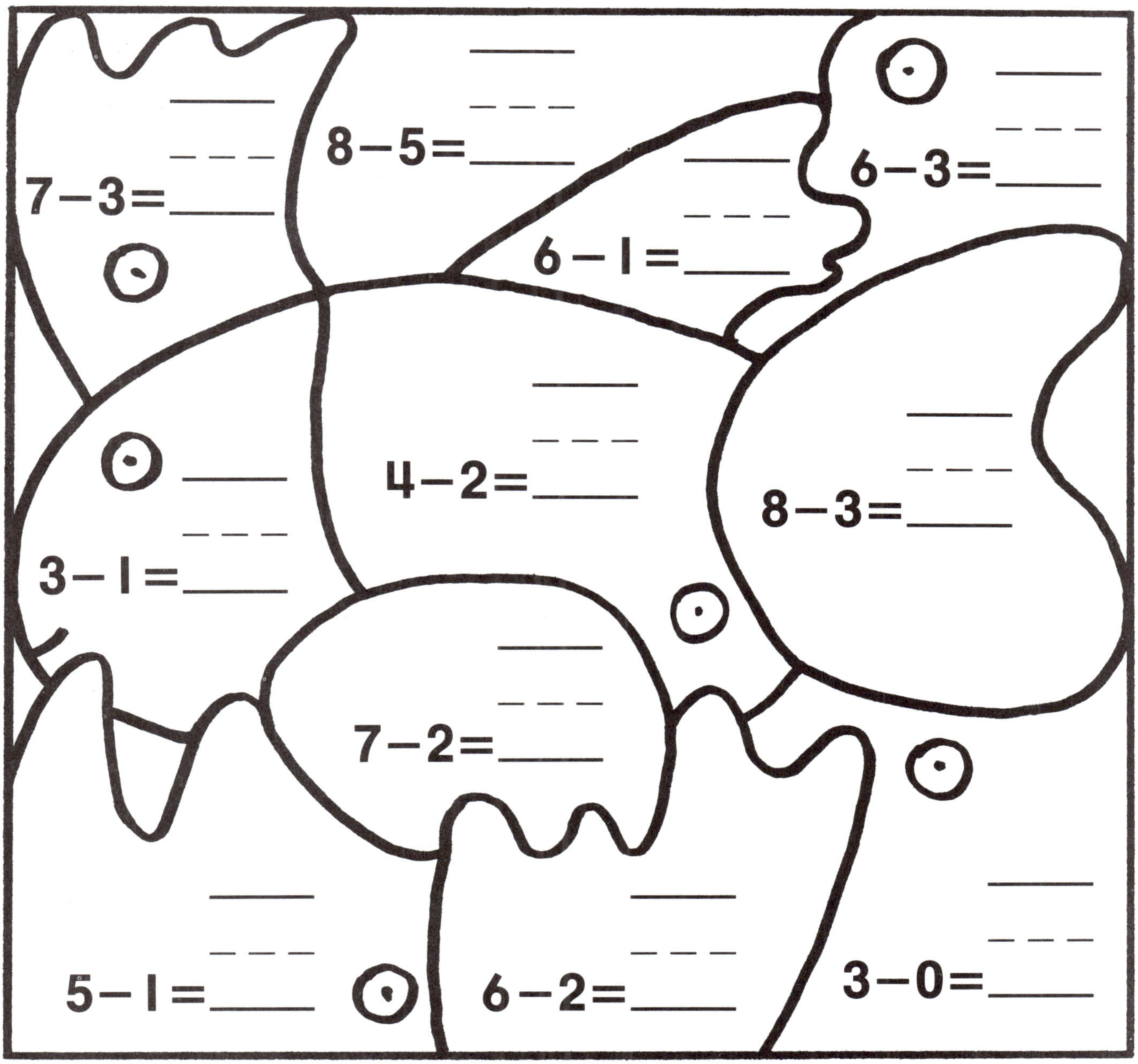

Fly Away

Write how many are left.

How many are left? 10 – 3 = 7

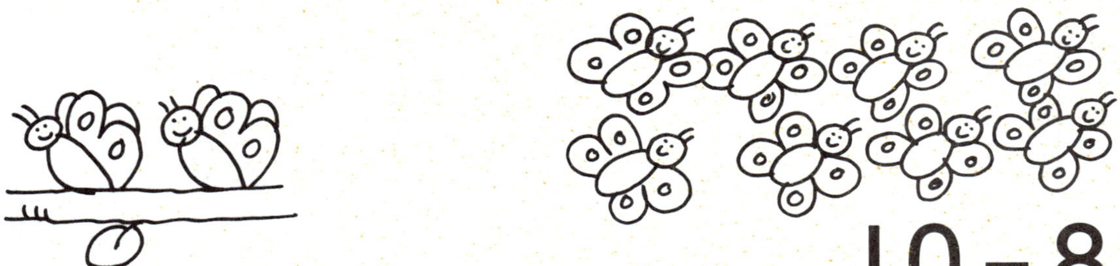

10 – 8 = ____

9 – 6 = ____

10 – 5 = ____

Use a sticker to answer.

9  – 7  = ____

Skill: To subtract numbers 0–10 from numbers to 10.

Clown Around

Write how many are left.

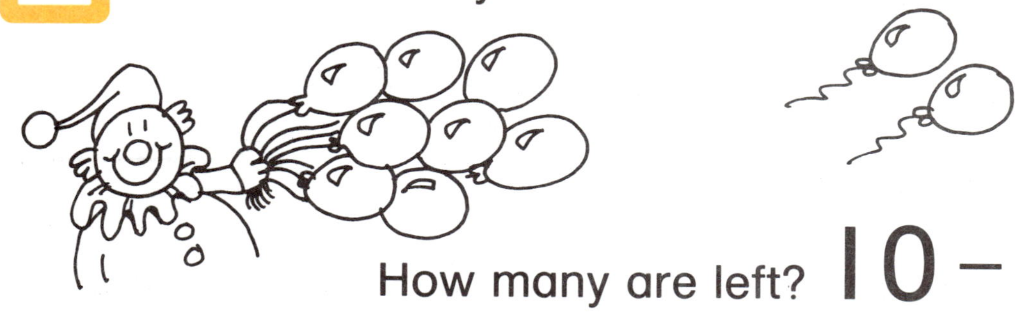

How many are left? 10 − 2 = 8

10 − 7 = ____

10 − 4 = ____

9 − 9 = ____

Use a sticker to answer.

10 − 8 = ____

Color Away

Subtract.

How many are left? $10 - 5 = 5$

How many are left? $10 - 6 = \_\_\_$

How many are left? $9 - 5 = \_\_\_$

How many are left? $8 - 6 = \_\_\_$

Use a sticker to answer.

What color is good to eat?

Skill: To subtract numbers 0–10 from numbers to 10.

Make Music

Cross out and subtract.

X out 7.

How many are left? $10 - 7 =$ 3

X out 7.

How many are left? $9 - 7 =$ ____

X out 8.

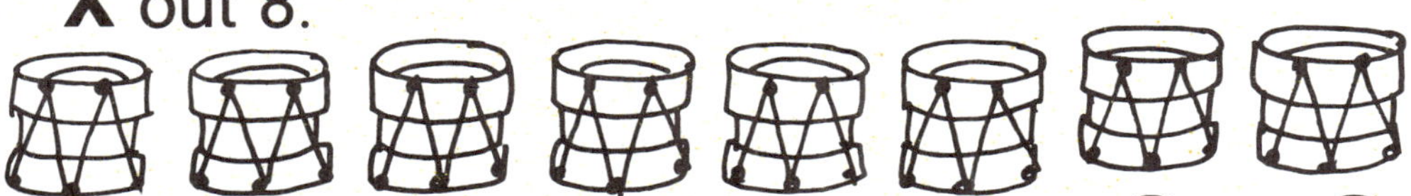

How many are left? $8 - 8 =$ ____

X out 5.

How many are left? $10 - 5 =$ ____

Use a sticker to answer.

10 small bells
rang so fine.
1 bell fell. Now there's

Skill: To subtract numbers 0–10 from numbers to 10.

Good and Fruity

Subtract.

$$\begin{array}{r} 10 \\ -\ 4 \\ \hline 6 \end{array}$$

How many are left? 6

$$\begin{array}{r} 9 \\ -\ 6 \\ \hline \end{array}$$

How many are left? ______

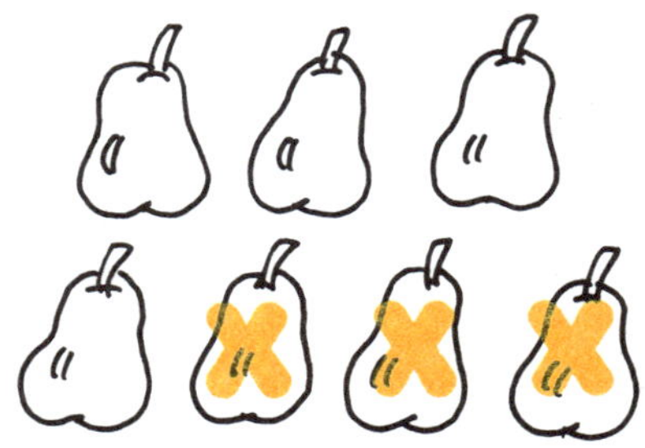

$$\begin{array}{r} 7 \\ -\ 3 \\ \hline \end{array}$$

How many are left? ______

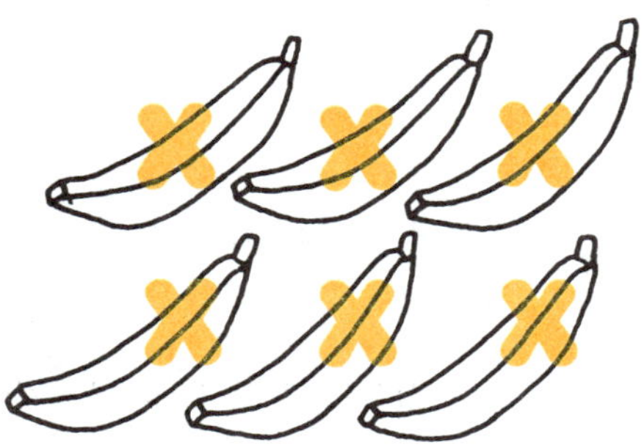

$$\begin{array}{r} 6 \\ -\ 6 \\ \hline \end{array}$$

How many are left? ______

$$\begin{array}{r} 10 \\ -\ 2 \\ \hline \end{array}$$

How many are left? ______

Use a sticker to answer.

What has a mouth but cannot talk?

Skill: To subtract numbers through 10 in vertical form.

Go Fish

Cross out and subtract.

$$\begin{array}{r} 9 \\ -\ 4 \\ \hline 5 \end{array}$$

How many are left? ____

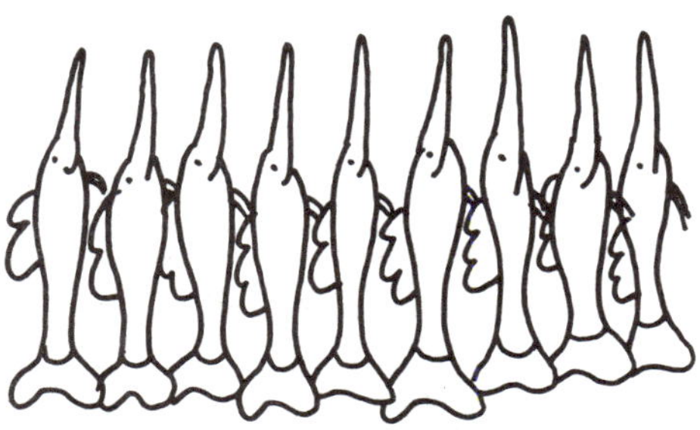

$$\begin{array}{r} 9 \\ -\ 8 \\ \hline \end{array}$$

How many are left? ____

$$\begin{array}{r} 10 \\ -\ 6 \\ \hline \end{array}$$

How many are left? ____

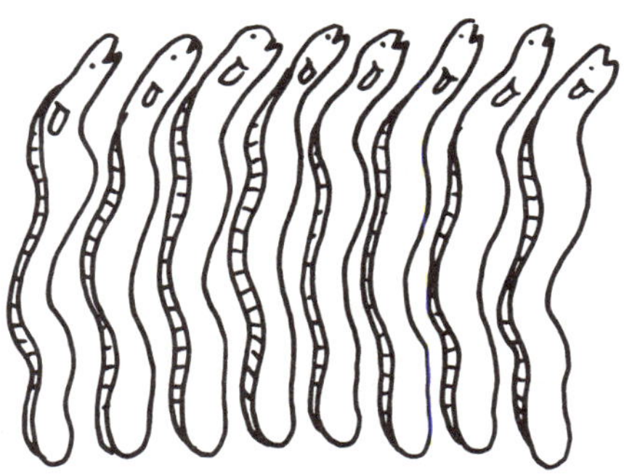

$$\begin{array}{r} 8 \\ -\ 5 \\ \hline \end{array}$$

How many are left? ____

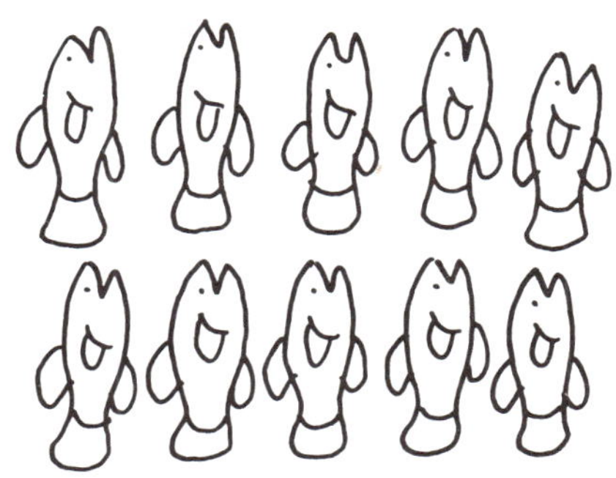

$$\begin{array}{r} 10 \\ -\ 7 \\ \hline \end{array}$$

How many are left? ____

Use a sticker to answer.

What runs but never gets tired?

Skill: To subtract numbers through 10 in vertical form.

Shape Up

Subtract.

Find the stickers to check your answers.

$$\begin{array}{r} 5 \\ -\ 2 \\ \hline 3 \end{array}$$

$$\begin{array}{r} 7 \\ -\ 3 \\ \hline \end{array}$$

$$\begin{array}{r} 9 \\ -\ 7 \\ \hline \end{array}$$

$$\begin{array}{r} 10 \\ -\ 5 \\ \hline \end{array}$$

$$\begin{array}{r} 6 \\ -\ 0 \\ \hline \end{array}$$

$$\begin{array}{r} 10 \\ -\ 2 \\ \hline \end{array}$$

Skill: To practice subtraction facts through 10.

Check It Out

Subtract. Then add to check.

$5 - 3 = 2$ $\qquad$ $2 + 3 = 5$

$9 - 5 = \_\_\_$ $\qquad$ $4 + 5 = \_\_\_$

$6 - 4 = \_\_\_$ $\qquad$ $2 + 4 = \_\_\_$

$10 - 3 = \_\_\_$ $\qquad$ $7 + 3 = \_\_\_$

Use a sticker to answer.

What has no legs but can jump?

Sticker Stories

Find the sticker that shows the story.

Write the answer.

5 

4 run away.

How many are left? ____

$5 - 4 =$ ____

---

9 

3 go out.

How many are left? ____

$9 - 3 =$ ____

---

10 

7 get picked.

How many are left? ____

$10 - 7 =$ ____

Skill: To solve subtract story problems and to find matching pictorial representations of them.

Many Pennies

Find the penny sticker.

Subtract pennies.
How much is left?

A penny is 1¢

I have 

I buy 

10¢ − 8¢ = ____¢

I have 

I buy 

9¢ − 5¢ = ____¢

I have 

I buy 

8¢ − 4¢ = ____¢

I have 

I buy 

10¢ − 10¢ = ____¢

Flower Power

Subtract.

$$\begin{array}{r} 6 \\ -\ 2 \\ \hline \end{array} \qquad \begin{array}{r} 8 \\ -\ 5 \\ \hline \end{array} \qquad \begin{array}{r} 10 \\ -\ 8 \\ \hline \end{array}$$

$$\begin{array}{r} 5 \\ -\ 5 \\ \hline \end{array} \qquad \begin{array}{r} 7 \\ -\ 3 \\ \hline \end{array} \qquad \begin{array}{r} 9 \\ -\ 1 \\ \hline \end{array}$$

$$\begin{array}{r} 8 \\ -\ 4 \\ \hline \end{array} \qquad \begin{array}{r} 7 \\ -\ 0 \\ \hline \end{array} \qquad \begin{array}{r} 10 \\ -\ 6 \\ \hline \end{array}$$

 Skill: To review subtraction facts to 10.

Practice Test

I can subtract across.

7 − 5 = 2

- ○ 4
- ○ 3
- ● 2

Subtract.

A. 5 − 1 = ___

- ○ 6
- ○ 5
- ○ 4

E. 7 − 3 = ___

- ○ 5
- ○ 4
- ○ 3

B. 10 − 9 = ___

- ○ 3
- ○ 2
- ○ 1

F. 9 − 6 = ___

- ○ 4
- ○ 3
- ○ 2

C. 8 − 4 = ___

- ○ 4
- ○ 3
- ○ 2

G. 10 − 7 = ___

- ○ 3
- ○ 2
- ○ 1

D. 6 − 6 = ___

- ○ 2
- ○ 1
- ○ 0

H. 9 − 4 = ___

- ○ 6
- ○ 5
- ○ 4

Skill: To test subtraction facts in horizontal form.

Practice Test

I can subtract down.

$$\begin{array}{r} 8 \\ -3 \\ \hline 5 \end{array}$$

- ○ 6
- ● 5
- ○ 4

Subtract.

A. $\begin{array}{r} 7 \\ -4 \\ \hline \end{array}$

- ○ 4
- ○ 3
- ○ 2

B. $\begin{array}{r} 6 \\ -1 \\ \hline \end{array}$

- ○ 7
- ○ 6
- ○ 5

C. $\begin{array}{r} 10 \\ -\ 6 \\ \hline \end{array}$

- ○ 6
- ○ 5
- ○ 4

D. $\begin{array}{r} 8 \\ -3 \\ \hline \end{array}$

- ○ 6
- ○ 5
- ○ 4

E. $\begin{array}{r} 9 \\ -5 \\ \hline \end{array}$

- ○ 5
- ○ 4
- ○ 3

F. $\begin{array}{r} 8 \\ -7 \\ \hline \end{array}$

- ○ 2
- ○ 1
- ○ 0

G. $\begin{array}{r} 9 \\ -9 \\ \hline \end{array}$

- ○ 2
- ○ 1
- ○ 0

H. $\begin{array}{r} 10 \\ -\ 5 \\ \hline \end{array}$

- ○ 4
- ○ 5
- ○ 6

Skill: To test subtraction facts in vertical form.

page 23

page 13

page 11

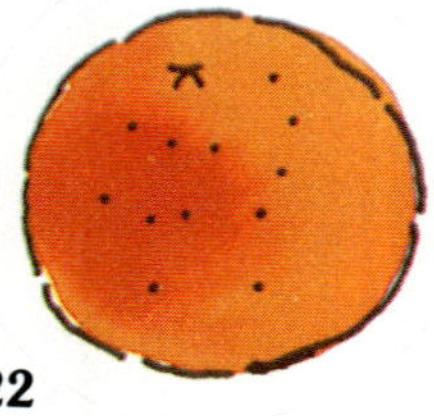

page 22

page 29

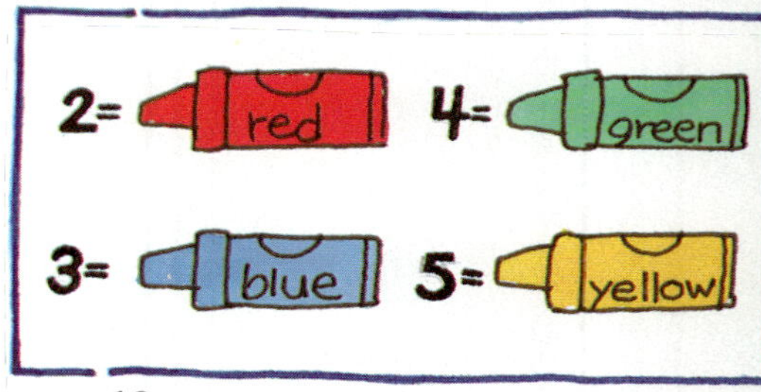

page 19

page 31

page 10

page 12

page 32

page 5

page 26

page 24

page 14